The Young Scientist Investigates

Electricity and Magnetism

by
Terry Jennings

CHILDRENS PRESS ®
CHICAGO

Illustrated by
Norma Burgin
Stove Crisp
Karen Dawes
Chris Molan
Tony Morris
Tudor Artists

Library of Congress Cataloging-in-Publication Data

Jennings, Terry J.
　Electricity and magnetism.
　(The Young scientist investigates)
　Reprint. Originally published: London : Oxford University
Press, 1982.
　Includes index.
　Summary: An introduction to magnetism, magnets, compasses,
batteries, and electricity. Includes study questions, activities, and
experiments.
　1. Electricity—Juvenile literature. 2. Magnetism—Juvenile
literature. [1. Electricity. 2. Magnetism.] I. Title. II. Series:
Jennings, Terry J. Young scientist investigates.
QC527.2.J46　1989　　　537　　　88-36215
ISBN 0-516-08437-2

North American edition published in 1989 by
Childrens Press®, Inc.

Printed in the United States of America
1 2 3 4 5 6 7 8 9 10 R 98 97 96 95 94 93 92 91 90 89

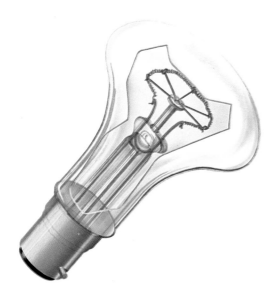

The Young Scientist Investigates

Electricity and Magnetism

Contents

Early magnets

More than 2,000 years ago, the Greeks found a strange stone. If a piece of the stone was hung from the end of a string, the stone always pointed in the same direction. The people who lived in the place where the stone was found were called Magnetes. It is from the name Magnetes that today's word *magnet* comes. The stone was a natural magnet.

Chinese travellers also used the same kind of stone. The stone could be used to guide people who were going on long journeys. Because of the help these stones gave to travellers, they were called lodestones. *Lodestone* means "leading stone." Lodestone is a natural magnet.

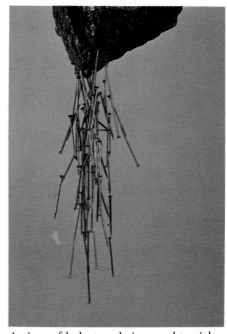

A piece of lodestone being used to pick up steel pins

Most of the magnets we use today are made of iron or steel. Some magnets are straight. They are called bar magnets. Other magnets are shaped like a horseshoe. Some magnets are round, while some are shaped like a pencil. These magnets are made in factories. Small nails and pins stick to these magnets as if they were glued.

Some different shaped magnets

3

Uses of magnets

Magnets are fun to play with. There are also magnets in electric train sets and lots of toys. But there are many other important uses for magnets. There are magnets built into a refrigerator door. These magnets help to keep the door shut.

Magnets help to make radio and television sets work. There are magnets in telephones and microphones. There are magnets in electric motors. Vacuum cleaners, refrigerators, freezers and washing machines are just some of the things in the home that are driven by electric motors. They all have magnets in them to make them work.

On these two pages are some things that have magnets in them to make them work.

4

Magnets help to make electricity at the power plant. The machines that make electricity have magnets in them. There is a magnet in a compass. So magnets help ship captains, airplane pilots and explorers to find their way.

Magnets are used to separate different kinds of metal at the scrap yard. Some cranes have a magnet on them instead of a hook. These cranes can be used to pick up big pieces of iron and steel. The magnet on a crane can also be used to drop a heavy steel ball to break things up.

A magnet on a crane being used to pick up pieces of steel

Compasses

If you tie a piece of string to the middle of a bar magnet, it will swing around. The magnet will swing around until one end points to the north magnetic pole of the earth, which is near the North Pole. The end of the magnet that always points north is called its north pole. The other end always points south. This end is called the magnet's south pole.

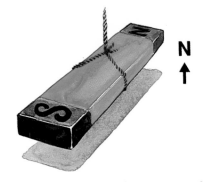

A bar magnet swinging from a piece of string

A pocket compass

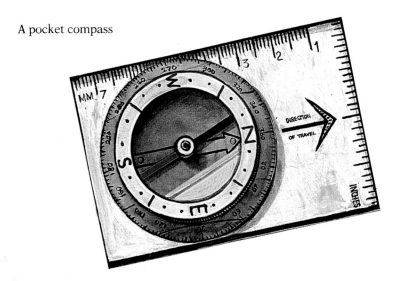

An early ship's compass

The pictures show compasses. A compass is used for finding the way. Ship captains, airplane pilots and explorers use a compass to find the way. The needle of a compass is a small, thin magnet. The compass needle always points north. A magnet swings so that it points north because the earth itself is like a big magnet.

A modern ship's compass

The earth goes round and round. It behaves as if it had a big magnet along its center. Small magnets are pulled toward the earth's big magnet. No one knows why or how the earth came to behave like a big magnet.

Attracting and repelling

Magnets will pull some things toward them. We say the things have been *attracted* to the magnet. This property is called *magnetism*. Magnets attract things made of metal, usually iron or steel. Magnets also attract things made of the metals cobalt and nickel.

The pull of a magnet is strongest at its ends or poles. One magnet sometimes attracts another magnet. If the north pole of one magnet is pointed towards the south pole of another, they will attract each other. But if you point the south pole of one magnet towards the south pole of another, they will not attract each other. The two magnets push each other away. We say the two magnets *repel* each other.

In the same way, the north pole of one magnet will repel the north pole of another. So two south poles repel each other. Two north poles repel each other. But a north and a south pole attract each other. We say, "Like poles repel; unlike poles attract."

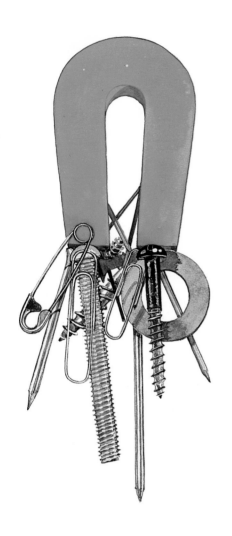

Bar magnets attracting each other (above) and repelling each other (below)

Making magnets

One way to make a magnet is to use a piece of steel. A sewing needle or a steel knitting needle will do. All you have to do is to stroke the needle with one end of a magnet. You must always stroke the needle with the same end of the magnet and in the same direction. The more times you stroke the needle, the more powerful a magnet it will be. Your magnet will stay a magnet as long as you use it carefully. But if you drop your magnet or hit it hard, it will not be as strong. It may even stop being a magnet altogether. All magnets can be weakened if you drop them or hit them hard.

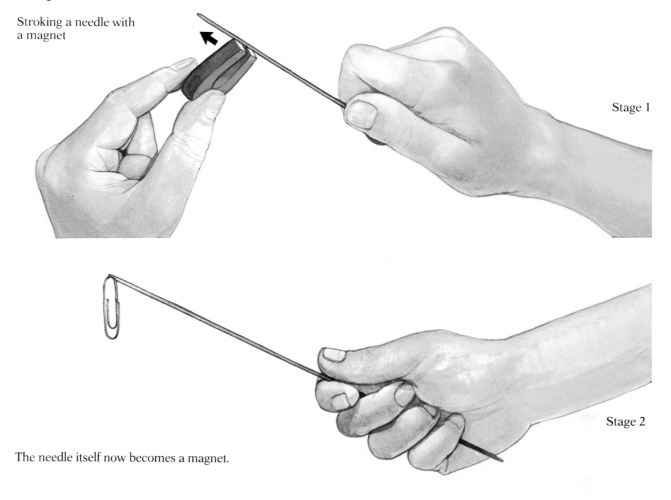

Stroking a needle with a magnet

Stage 1

Stage 2

The needle itself now becomes a magnet.

Electromagnets

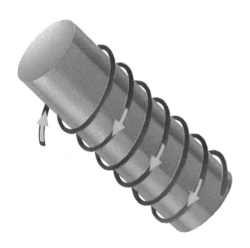

There is another kind of magnet. It is not a magnet all the time. It is a piece of iron with a coil of covered wire around it. When electricity passes through the wire the iron becomes a magnet. When the electricity is switched off, the iron is no longer a magnet. This kind of magnet is called an electromagnet. An electromagnet is often used to help smash up old cars and machinery. It can also be used on a crane to pick up iron and steel from heaps of other metals. The crane in the picture on page 5 has an electromagnet on it.

An electromagnet on a crane being used to pick up scrap iron and steel

Do you remember?

(Look for the answers in the part of the book you have just been reading if you do not know them.)

1   Where does today's word "magnet" come from?

2   What are today's magnets made from?

3   Name six things found in the home that have magnets in them.

4   Which direction does a magnet hung on a string point to?

5   What is the needle of a compass made of?

6   Who uses a compass?

7   What metals are the things made of that are attracted to a magnet?

8   What happens if two magnets are placed so that the north pole of one is close to the south pole of the other?

9   What happens if two magnets are placed so that the south pole of one is close to the south pole of the other?

10   What happens if two magnets are placed so that the north pole of one is close to the north pole of the other?

11   How could you turn a sewing needle into a magnet?

12   How could you weaken a magnet?

13   What is an electromagnet made from?

14   What happens if you turn off the electricity to an electromagnet?

15   What are electromagnets used for?

Things to do

1   **Write a story.** Pretend that you lived hundreds of years ago in the part of China or Greece where rock that was a natural magnet was found. Write a story about how you discovered this strange rock and what you did with it.

2   **Hold a magnet against a number of different objects.** Try as many different things as you can find. Make two lists like this:

Magnets attract these	Magnets do not attract these
Pins	Plastic comb
Nails	Bottle
	Pencil

Now look at your list of objects that are attracted by a magnet. What are they all made of? Are they all the same color?

3 A boat race. Stroke a small nail with a magnet to make the nail become a magnet itself. Stick the nail into a cork. Float the cork on a bowl of water. Make your nail boat move by holding a magnet near it. Make your nail boat sail in all directions.

Have a nail boat race with a friend.

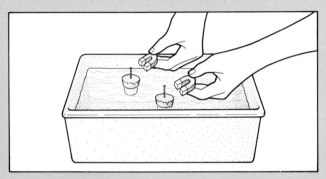

4 A traffic game. Find a sheet of cardboard. Lay it on two blocks of wood or two piles of books. Leave enough room to move your hand around underneath the cardboard.

Find a small toy car that has steel in it. Place it on the cardboard. Now with your magnet held underneath the cardboard, you can make your car move.

Mark out some roads on the sheet of cardboard and make road signs out of

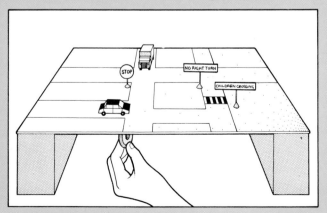

index cards. Stand them up with pieces of clay. Play a traffic game with a friend who also has a magnet and a toy car or truck. Make the traffic obey the road signs you have made.

5 What objects can be magnetized? Use a magnet to magnetize some of the things in the home that are made of steel. You might try making a screwdriver, penknife or pair of scissors into a magnet.

6 Make a compass. Cut a strip of cloth about 1½ inches wide and 6 inches long. Make a hole in the center of each end of the piece of cloth. Take a short piece of string (about 12 inches long). Tie one end into each hole in the cloth. You have now made a sling.

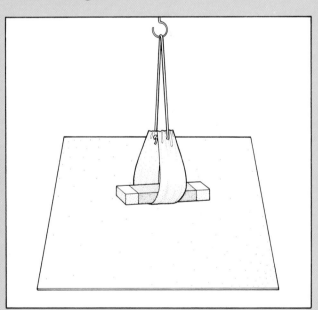

Place a bar magnet in the sling. Hang the sling up by the loop of string. Which way does the magnet point? Push the magnet slightly. What happens? Does the magnet always point the same way?

If you already know which way is north, you now know which end of your magnet is the north pole. You can then mark it with a spot of paint. You have now made a compass. What happens if you hold another magnet near to your compass? What happens if you hold a piece of iron or steel near to your compass? Can you think why the boxes of compasses are never made of iron or steel?

Another way to make a compass is by using a steel sewing needle. Make the needle into a magnet by stroking it with a bar magnet. Hold the needle over some iron filings to make sure that it has been magnetized.

Cut a thin circle of cork from a bottle stopper. Float the cork on a saucer of water. When the water is still, rest the needle on top of the cork. The needle will then swing so that one end points to the north and the other to the south.

7 Direction. Take a compass into the yard or playground. Set the compass so that the north on the scale and the point of the needle are both pointing in the same direction. Make lists of the buildings and places that are north, east, south and west of where you are standing.

8 Make a magnet "float" in the air. For this you need two bar magnets that are the same size. Lay one magnet on a piece of wood or cardboard. Use a skewer or a nail to make small holes around the magnet but 1/16 inch away from it. Make the holes about 1/3 inch apart.

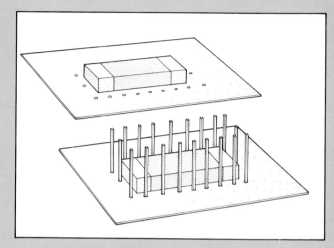

Glue a matchstick into each hole.

Now lay the other magnet over the first and inside the fence of matchsticks. The magnets should be placed so that the north pole of one is next to the north pole of the other. How can you find out which end is which if they are not marked? (Remember what you read on page 7!)

What do you notice about the top magnet? Push down on it gently. What happens? Can you think of any way in which magnets like this could be used instead of springs?

If you have some toy bricks, you could make a fence of these around the magnet instead of using matchsticks.

9 Do the Indian rope trick with a magnet. In the Indian rope trick, a rope is made to stand up straight without anyone holding it.

Find a small block of wood. Stick a tack in the middle of it. Take a piece of thread about 12 inches long and tie one end to the tack. Tie the other end of the thread to a paper clip.

Pull the paper clip so that the thread (your piece of "rope") is standing up straight. Then hold a magnet just above the paper clip and let the clip go. Can you make the "rope" stay upright?

Can you find any way of fixing the magnet so that it can be left just above the paper clip?

10 Play a fishing game with magnets. Cut out some fish shapes from index cards. Fix a paper clip to each one. Make sure that the paper clips are made of steel and can be picked up with a magnet.

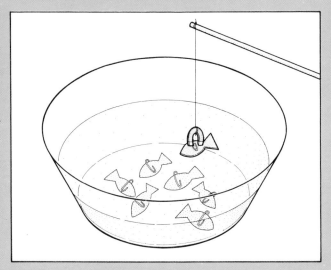

Put your fishes in a bowl of water. Take a short stick and tie some string to the end of it. Hang a magnet from the end of the string.

Have a fishing match with a friend. See how many fish you can each catch from the bowl in one minute.

11 Make patterns with a magnet. Lay a bar magnet on the table. Take two books that are the same thickness. Place one book on each side of the magnet. Lay a thin sheet of white poster board over the books and magnet. Sprinkle iron filings over the board. Now tap the board gently.

What happens to the iron filings? Draw the pattern they make.

Now see what patterns you can make using a horseshoe magnet.

Do you get a different pattern if you put two magnets near to each other under the card? Try two magnets with their north poles side by side. Then try two magnets with their north and south poles side by side.

12 Funny Faces. Draw a picture of a man's face on a sheet of poster board. Put some iron filings on the man's face to give him a beard. Carefully slide a magnet under the picture. Can you make the man's whiskers move? Can you move his beard to another part of his face?

Magnet

Iron filings

13 Use your imagination. Imagine that one day a giant who was 75 feet high made a huge magnet. What do you think the giant might have done with the magnet? Was he a good giant or a bad one? Write a story about the giant and the magnet.

Electricity

We use electricity in hundreds of ways. Almost everyone uses electric light. Many people cook by electricity and use it to warm their homes. We also use electricity to keep our food cool and fresh. Most refrigerators operate on electricity, and so do freezers.

Electricity can also be used to give us hot water. Many people use electricity to clean their homes. The vacuum cleaner has an electric motor that sucks dust and dirt from our carpets. We use electricity to press our clothes and to dry our hair.

Electricity can even amuse us and teach us. Radio and television sets use electricity. So do tape recorders and record players. The projectors that show films at the movies are driven by electricity. Electricity drives some trains. It also drives some cars, vans and buses.

A store that sells electrical goods

A refrigerated milk truck

An electric subway train

How many things that use electricity can you see in the picture opposite?

15

Power plants

The electricity we use in our homes, shops and factories is made in a power plant. The power plant in this picture burns coal. Trains or boats bring the coal to the power plant. Many power plants are built by rivers or canals so that boats can bring the coal. Other power plants have railway sidings to which special trains carry the coal. Inside the power plant the coal is burned in large furnaces. The boilers heat water and turn it into steam.

Steam can push very hard. Steam in a saucepan or kettle can push up the lid. In the power plant the steam pushes hard against a big wheel. The big wheel is called a turbine. The steam pushes the turbine round and round. As the turbine goes round it turns a machine called a generator. As the generator turns it makes electricity. Some kinds of power plants burn oil or gas to heat the water. Oil and gas are cleaner fuels than coal. They do not make a lot of smoke. And so the air stays cleaner.

A coal-burning power plant

The turbines (painted blue) and the generators (painted yellow) inside a power plant

An oil-burning power plant

A hydroelectric power plant. What do you think the big pipes are for?

An atomic or nuclear power plant

Hundreds of years ago people used fast rivers to turn their mills to grind wheat. Some of these water mills can still be seen today. Nowadays electricity is sometimes made by using water to turn a big water wheel. The big water wheel is a turbine. Water from a fast flowing river on a hillside or mountain is led by pipes into a power plant. The water turns the turbines. The turbines turn the generators that make electricity.

This kind of power plant is called a hydroelectric power plant. Hydroelectric power plants are built where there are mountains and fast rivers. There are some hydroelectric power plants in the United States and Canada.

The picture on the left shows an atomic or nuclear power plant. An atomic power plant does not need a fast river. It does not need coal, oil or gas. An atomic power plant uses a little piece of a mineral called uranium. Uranium comes from some kinds of rock. Uranium gives off atomic energy. Atomic energy from the uranium heats the water in the power plant. The water turns into steam to make electricity.

Wires and cables

On the walls inside the power plant there are many meters. A meter looks like a clock with only one hand. Some meters show how much electricity is being made. Other meters show how much electricity is being used in the different towns and villages.

The control room inside a power plant

From the power plant the electricity goes along wires. Wires are a way of moving electricity from one place to another. Power plants send electricity along thick wires that are put on high towers called pylons. Near towns the thick wires branch into many smaller wires. The smaller wires are carried on poles. The wires go along the streets from pole to pole. Often you cannot see the electricity wires in towns because they are put together inside special covers that protect them. These are called cables. The cables are buried under the streets.

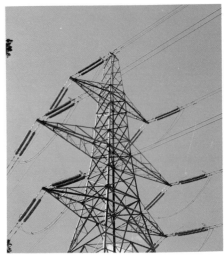
The wires on a pylon

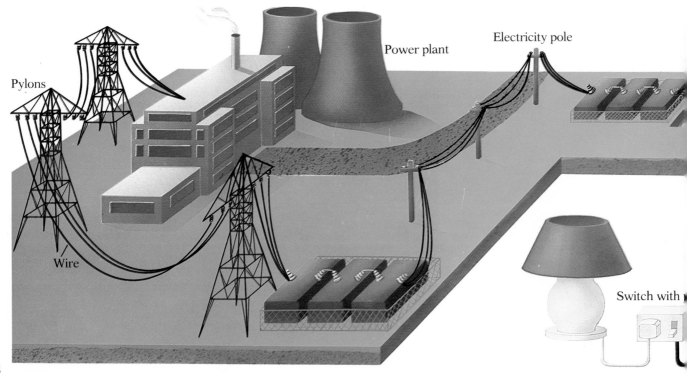
Pylons
Wire
Power plant
Electricity pole
Switch with

Electricity in the home

An electricity meter

From the big cables in the road or under the street, a smaller cable goes into each house. The cable goes into the house to a little box. The box has dials behind a glass window. The box is the electricity meter. It measures how much electricity you use. This picture shows one kind of meter. Someone comes to your house to read the meter to see how much electricity you have used. We pay for the electricity that we use.

Close by the meter are the main switches. They turn off all the electricity to your home. These switches can be used to turn off the electricity if something goes wrong.

Your house may have circuit breakers or fuses. All the electricity you use goes through the circuit breakers or the fuses. These devices cut off the flow of electricity if something goes wrong with an appliance that works by electricity and the circuit becomes overloaded, or overheated. Circuit breakers are switches that turn the electricity off. Fuses contain thin pieces of wire that melt easily. If a circuit becomes overloaded, the fuse heats up and melts, cutting off the electricity. Some electrical plugs have a fuse.

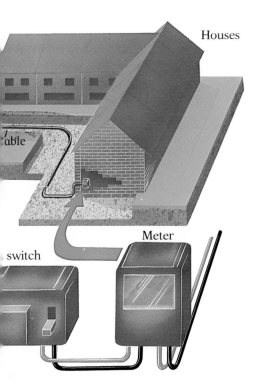

Houses

cable

switch

Meter

The fuse box in a house

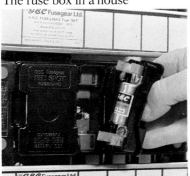

The fuse inside an electrical plug

It is *not* safe for children to touch fuses.

Electric lights

Most people use electric lights at home. We also light up the streets and shops and factories with electric lights. Nearly all of these lights come from an electric light bulb. The bulb is made of glass. Inside the bulb is a very thin wire. When electricity passes through the wire it becomes very hot. The wire becomes so hot that it glows. The glowing wire gives us light. The bulb is filled with gas. The gas stops the thin wire from burning away too quickly. A flashlight bulb works in the same way as the large light bulbs we use in our houses.

Sometimes we use fluorescent tubes to give us light. The tube is filled with gas. The inside of the tube is coated with a special kind of paint. When the electricity is turned on, the gas in the tube makes the paint glow. The glowing paint gives us light.

Inside an electric light bulb

Using electric lights

A fluorescent light tube

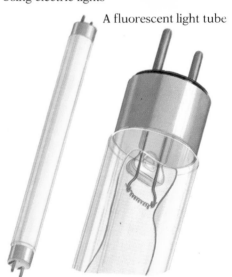

London at night

Safety first

When electricity is used carefully, it helps people. But electricity can be dangerous. It can start a fire and it can hurt people. Do not touch anything electrical if your hands are wet.

Electricity flows easily through some materials such as metals and water. These materials are called *conductors*. Other materials such as rubber and plastic do not carry electricity. These are called *insulators*. In electric cords, the metal wires are covered with rubber or plastic to keep the electricity inside. Electric cords inside a house should always be well covered with rubber or plastic. If the cover is torn, it can be very dangerous. A cord with a torn cover could injure you or start a fire.

How a damaged cord can be dangerous

The electric wires on pylons and poles are not covered. The wires do not usually touch anything that burns. But sometimes these wires are blown down in a bad storm. A fallen wire is very dangerous. Do not go near it, but ask a grown-up to call the police or the electric company.

Wires like this are dangerous because they are not covered.

Always dry your hands thoroughly before you touch anything electrical.

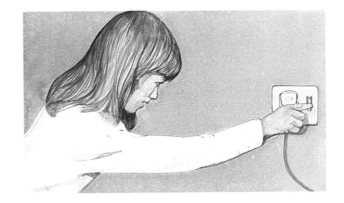

Batteries

We get most of our electricity in two ways. Much of our electricity comes from a generator. This is the electricity that is made at the power plant. Bicycles may also have a small generator or dynamo on them to make the lights work. We also get electricity from dry-cell batteries and storage batteries.

The dynamo on a bicycle

The picture shows you what a dry-cell battery looks like inside. The chemicals inside the battery make electricity. As the electricity is taken from the battery, the chemicals change. In the end the chemicals are used up and no more electricity is made. The battery can then be thrown away. Sometimes two batteries are joined together as in the picture below.

Inside a dry-cell battery

Another kind of battery is called a storage battery. After it has used its electricity, it can be recharged and used again. Cars and trucks carry storage batteries. As the electricity is used up, more is made by a generator attached to the engine. The battery stores the electricity until it is needed by the lights and starter motor of the car or truck.

A car battery

Two batteries joined together

There are several dry cells joined together inside this 9-volt battery.

Inside a 9-volt battery

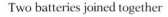

Electrical circuits

All batteries have two metal terminals. You can see the terminals on the batteries in the picture. Some batteries have only one terminal on the top. In these batteries, the bare metal at the bottom is the other terminal. When the battery is not used, no electricity flows. But if we put a wire from each terminal of the battery to each terminal on a lampholder, the electricity will flow. The electricity flows out of one terminal of the battery through the bulb, and back into the battery. The bulb lights up. We call the way in which electricity comes out of the battery, through the bulb and back into the battery, a *circuit*. A circuit is a circle with no breaks.

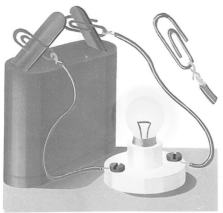

Electricity flows as long as the circuit is joined. A switch acts like a gate in the circuit. When the switch is turned on, the gate is closed and the circuit is closed. The electricity can then flow. When the switch is turned off, the gate is open and the circuit has a gap in it. The electricity cannot then flow. All electricity flows in a circuit. All the electrical things in your home work when the circuit is closed.

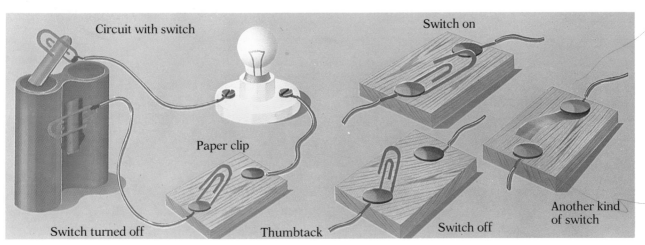

Circuit with switch

Switch on

Paper clip

Switch turned off

Thumbtack

Switch off

Another kind of switch

Static electricity

Electricity comes from a power plant in a continuous flow, or current. It is called current electricity. There is another kind of electricity called static electricity. A static electric charge builds up until it discharges all at once.

Sometimes you can see sparks of electricity when you comb your hair in a darkened room. If you rub a comb on your sleeve, you can use it to pick up little pieces of paper. The comb picks up the paper because of static electricity. If you rub a balloon against your sleeve, you can make it stick to a wall. As you take off your nylon clothes they often crackle because of static electricity on them. You can sometimes see sparks if you walk across a carpet and then touch a light switch.

When storm clouds get too full of static electricity they may spark to each other or to the ground. Lightning is just a big spark of static electricity. The electricity builds up in the cloud until it discharges. Lightning sparks are very powerful. They can damage trees and buildings, or start a fire.

Sparks of electricity may be seen when you comb your hair in a darkened room.

A comb rubbed on the sleeve will pick up small pieces of paper.

Lightning

Electric fish

An electric ray

An electric eel swimming

Another kind of electricity is found in the water. Some water creatures make a lot of electricity in their bodies. They make electricity in their muscles. Electric rays are fishes that live in the sea. They make a lot of electricity in the muscles of their bodies. The electric rays use the electricity to stun the other fishes they eat for food. They also drive off their enemies with electricity.

The electric eel is a fish that lives in rivers in South America. The electric eel makes a lot of electricity in its muscles. Electric eels live in very muddy water. The water is so dark that the electric eels cannot see. But they use electricity to find their way along. They use the electricity rather like a ship's radar. Electric eels also use electricity to kill their food.

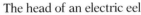
The head of an electric eel

Do you remember?

1   Make a list of 10 things that use electricity.

2   What is the coal, oil or gas used for in a power plant?

3   What is the machine called that turns and makes electricity?

4   What do we call a power plant that uses fast-flowing water to turn its turbines?

5   What does an atomic or nuclear power plant use instead of coal, oil or gas?

6   What does a meter look like?

7   What do the meters inside a power plant show?

8   What are cables?

9   What are the tall towers called on which big wires are put?

10 What does the electricity meter in your house show?

11 What are the switches near the electricity meter in your house used for?

12 What is a fuse and what does it do?

13 What is inside an electric light bulb and how does it work?

14 What is inside a fluorescent tube and how does it work?

15 Why is a cable with a torn cover dangerous?

16 What should you do if you find an electricity wire that has been blown down in a storm?

17 How does a flashlight battery make electricity?

18 What does the battery in a car or truck do?

19 What are battery terminals?

20 What do we call it when electricity flows from a battery, through a bulb and back to the battery?

21 What does a switch do?

22 What do we call the electricity that is made by rubbing things together?

23 What is lightning and how is it made?

24 What do electric rays and electric eels use electricity for?

Things to do

In this book we shall use batteries to give us electricity. They are perfectly safe to use. *Never* experiment with electricity from a wall outlet. It can give you a nasty shock or burn you. It could even kill you.

1 Make an electric switch. Find a small block of wood. Push a thumbtack a little way into the wood at one end. Wrap the bare end of a piece of covered wire around the thumbtack, and then push the tack farther into the wood.

Find a thin piece of metal. A paper clip or a narrow strip of aluminum foil will do. Attach the metal strip to another thumbtack as shown in the picture. Wrap the bare end of another piece of wire around this thumbtack, and then push the thumbtack into the wood. The metal strip should overlap the first thumbtack but not touch it.

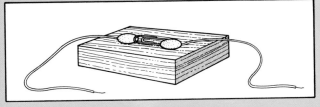

Find another piece of wire, and join up your switch to a bulb and battery, as shown in the picture below. You can now complete the circuit, so that electricity will flow, by holding the metal strip onto the thumbtack. When you let the metal strip go, the circuit is broken and the light is switched off.

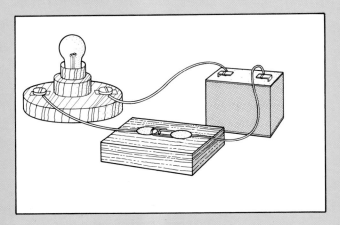

2 Make a working model lighthouse. You need a clean, empty liquid-detergent bottle. Pull the top off the bottle. Make a small hole in the side of the bottle near the bottom. Thread two pieces of covered wire about 20 inches long through the hole you have made, and out of the top of the bottle.

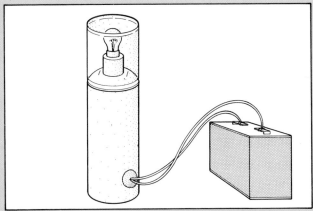

Plug the space around where the two wires go into the bottom of the bottle with clay. Then put a little dry sand or soil in the bottom of the bottle, to make the bottle less easy to knock over.

Ask a grown-up to cut off the top of the bottle until a lampholder just fits into the hole. Make a small cut in the top of the bottle on either side of the lampholder. Join one end of each wire to each terminal on the lampholder. Join the other two ends of the wires to the terminals on a battery. The bulb should light up.

Cover the lampholder with a small clear glass jar. A baby food jar is ideal. Paint the tower of your lighthouse. Most lighthouses are white but some have red and white stripes around them. Lighthouses also have a door and a small window in them.

Experiments to try

Your lighthouse would be easier to use if you had a switch to turn the light on and off. Can you make one?

Make a model harbor from cardboard or wood to go with your lighthouse.

3   Paint a picture of a power plant. Will it be a hydroelectric power plant, a coal-fired one, or one that uses gas, oil or atomic energy? Show in your picture how the water or fuel gets to the power plant.

4   Collect pictures of things that use electricity. Make your pictures into a wall-chart or scrapbook. Write a sentence or two about each picture.

5   How does a flashlight work? Ask permission to take a flashlight apart. Take it apart carefully. Draw a picture to show how the parts inside the flashlight are arranged.

8   Make a poster. The poster will tell people how to avoid having an accident with electricity. Make up words for your poster and paint a picture on it.

Never experiment with electricity from a wall outlet. It can give you a nasty shock or burn you. It could even kill you.

Do your experiments carefully. Write or draw what you have done and what happens. Say what you have learned. Compare your findings with those of your friends.

1   Does magnetism pull through different materials?

What you need: A magnet; some iron filings or small steel pins or paper clips; some index cards; a jam jar; various materials such as wood, rubber, leather, plastic and plastic wrap.

What you do: Spread some of the iron filings, pins or paper clips on an index card. Move the magnet underneath the card. What happens to the iron filings, pins or paper clips? Does magnetism pass through the card?

Now try the experiment with two cards placed one on top of the other. Does magnetism pass through the two cards? Try with more and more cards placed on top of each other. What happens?

Drop some iron filings, pins or paper clips into a jam jar. Will magnetism pass through the glass? What happens if you put water in the glass? Does magnetism pass through both the water and the glass?

Now try other materials such as wood, rubber, leather, plastic and plastic wrap. Will magnetism pass through these materials? If possible, try thick and thin pieces of each material.

Will magnetism pass through your hand? You may need a powerful magnet to find out.

3   How does an electromagnet work?

What you need: About 36 inches of thin covered wire; a large nail; 2 paper clips and a bicycle lamp battery; some small metal objects such as pins, small nails, iron filings or tacks.

What you do: Be sure that the two ends of the pieces of wire are bare. Wind your piece of wire neatly and evenly 15 or 20 times around the nail. Twist each end of the wire onto a paper clip.

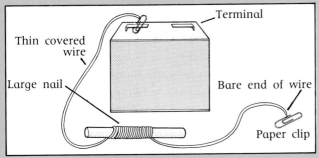

Fasten one paper clip onto one terminal of the battery. Place some steel pins, tacks, small nails or iron filings near to the nail. Now fasten the other paper clip onto the other terminal of the battery. Now that the electricity is flowing, what happens to the pins, tacks, small nails or iron filings? How many are picked up? What happens when you undo one of the paper clips?

Now wrap the wire twice as many times (30 or 40 times) around the nail. Connect the wires to the battery. How many pins, tacks, small nails or iron filings does the electromagnet pick up now? Is it more or less than before?

Make a list of the things your electromagnet will pick up. Make another list of the things the electromagnet will not pick up.

Try attaching a switch to your electromagnet. Make a drawing of the electromagnet.

What other things can you make an electromagnet with, apart from a large nail?

4   How can you make a simple circuit using a single-cell battery?

It is quite easy, as we have already seen, to join wires to a bicycle lamp battery or to any other battery that has two separate terminals. But how many ways can you find of making a simple circuit using one of the batteries in which the bare metal at the bottom is the second terminal?

What you need: A lampholder with bulb; a small screwdriver, a C or D battery; some pieces of wire; some paper clips; a penknife; a pair of all-metal scissors; some strips of aluminum foil; some clay; some clear tape.

What you do: Find a way of joining a piece of wire to the bottom of the battery using clay or clear tape. Fasten the other end of the wire to one of the terminals on the lampholder. Now fasten another piece of wire to the terminal on top of the battery and join it to the other terminal on the lampholder. Does the bulb light?

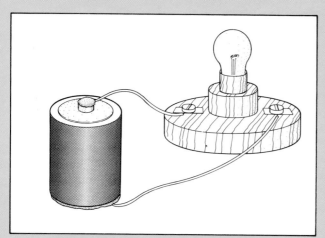

Now take the pieces of wire away. Can you get the bulb to light using only pieces of aluminum foil? Can you make it light using only paper clips? Now try the penknife or the pair of scissors.

What have you learned about wire, aluminum foil, paper clips, penknives and scissors?

5   Which materials are conductors and which are insulators?

What you need: A battery; two paper clips; a lampholder with a bulb; three pieces of wire; a small block of wood; two thumbtacks; a screwdriver; a key; an eraser; a pencil; a thick rubber band; a piece of string; a strip of cardboard and a selection of other materials.

What you do: Push the two thumbtacks into the block of wood. The tacks should be about 1 inch apart. Join up these two thumbtacks to the battery and lampholder as shown in the picture.

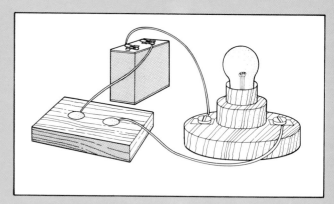

See that all the connections are tight. If you now lay the blade of the screwdriver across the two thumbtacks, the bulb should light.

Now lay the handle of the screwdriver across the two thumbtacks. Does the bulb light? Now try the other materials.

Materials that let electricity pass through them and make the bulb light are called *conductors.*

Materials that do not let electricity pass through them so that the bulb does not light are called *insulators.*

Make two lists, one of materials that are conductors of electricity and the other of materials that are insulators.

6   What things can be picked up using static electricity?

What you need: A plastic comb or a plastic ballpoint pen; a piece of cotton cloth; scraps of materials of various kinds such as

newspaper, tissue paper, cotton balls, feathers, pieces of plastic wrap and aluminum foil, grains of salt, sand or sugar, and tiny pieces of cork.

What you do: Make sure that everything is very dry. Then rub the comb or pen with the cotton cloth. Hold the comb or pen over some tiny pieces of newspaper. Are they picked up?

Now rub the comb or pen with the cloth again. Try to pick up little pieces of one of the other materials. Try each of the other materials in turn.

Turn the water on so that there is a steady trickle. Rub the comb or pen on the cotton cloth. Hold the comb or pen near to the trickle of water. What happens? Is the water attracted to the comb or pen?

Make two lists. In one list put things that are attracted to the comb or pen. In the other list put things that are not attracted to the comb or pen.

Now do the experiment again. Instead of using a comb or a pen try something else. You could use the handle of an old toothbrush, a plastic or wooden ruler, an eraser, a pencil, a piece of wood, and so on. Rub each one in turn with the piece of cotton cloth. See what, if anything, is picked up by each of these things.

7 Experiments with balloons and static electricity

What you need: Two balloons; some thread; a wool scarf or sweater; some clear tape.

What you do: Blow up one of the balloons. Tie the top with thread. Rub the balloon on the wool scarf or sweater. See if the balloon will "stick" to the wall. How long will the balloon stay there? If you rub the balloon more does it stay on the wall longer?

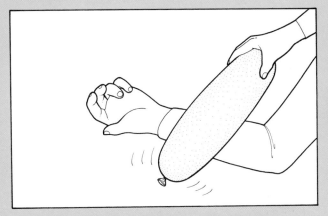

Rub the balloon on the scarf or sweater again and see what other things it will stick to. Try it on doors, curtains, cupboards, the ceiling, and so on.

Rub the balloon on the scarf or sweater. Hold it near small pieces of paper, dry grains of sand, salt or sugar. Are these things attracted to the balloon?

Now blow up the other balloon. Tie the top with thread. Rub both balloons on the scarf or sweater. Use clear tape to hang the two balloons side by side from a shelf or doorway. Put the two sides of the balloons you rubbed next to each other. What do you notice?

Glossary

Here are the meanings of some words that you might have met for the first time in this book.

Atomic energy: the energy that comes from a mineral called uranium, which is found in some rocks.

Attract: to draw something towards you.

Battery: a store of electricity.

Cable: a thick covered wire that carries electricity.

Circuit: the complete path, or way around, of electricity from a battery or generator, through a lamp or some other piece of electrical equipment, and back to the battery or generator.

Compass: an instrument that has a needle that is a magnet and always points to the north.

Conductor: a material through which electricity can flow.

Current: the flow of electricity.

Electromagnet: a magnet that works only while electricity is flowing through it.

Fuse: a thin piece of wire that melts easily and breaks an electrical circuit when something goes wrong.

Generator: a machine that makes electricity. It is often called a dynamo.

Hydroelectric power plant: a power plant that uses the force of running water to turn its generators.

Insulator: a substance through which electricity will not flow.

Magnetize: to turn something into a magnet.

Meter: an instrument that measures something, for example, an electricity meter measures how much electricity is being made or used.

Minerals: the chemical substances that make up rocks.

Poles (of a magnet): the places near the ends of a magnet from which the main magnetic pull seems to come.

Power plant: the large building where electricity is made.

Pylon: a tall tower on which electricity cables are hung.

Repel: to drive something away.

Static electricity: the electricity that is made by rubbing things together.

Terminal: a fastening by which a wire can be fixed to a battery, a switch, a lampholder or some other piece of electrical equipment.

Turbine: a large wheel or fan that is turned by running water or steam and that drives the generators that make electricity.

Acknowledgments

The publishers would like to thank the following for permission to reproduce transparencies:

Ardea London p. 25 (top); Barnaby's Picture Library p. 9, p. 20 (center), p. 21; Bruch Coleman Ltd p. 24, p. 25 (bottom) and front cover; Electricity Council p. 16 (top), p. 18 (top), p. 19 (top); IGS p. 2; T. Jennings p. 17 (bottom), p. 18 (bottom and back cover), p. 22 (bottom), p. 23 (top); London Transport p. 14 (bottom); Milk Marketing Board p. 14 (center); Picturepoint p. 4, p. 5; A Smith p. 22 (top); A Souster p. 14 (top); ZEFA p. 17 (top), p. 20 (bottom).

Illustrated by Norma Burgin, Stove Crisp, Karen Daws, Chris Molan, Tony Morris, and Tudor Artists

Index